W9-AOU-881
3 1192 00751 6618

GOBS OF GOO

by Vicki Cobb

illustrated by Brian Schatell

HarperCollinsPublishers

CONTENTS

Gobs of Goo

Library of Congress Cataloging in Publication Data
Cobb, Vicki.
Gobs of goo.
Summary: Describes various types of sticky substances and shows how they are made and used in everyday life.
1. Sugars—Juvenile literature.
2. Polysaccharides—Juvenile literature.
3. Lipids—Juvenile literature. 4. Thermoplastics —Juvenile literature. [1. Sugar. 2. Starch.
3. Oils and fats. 4. Plastics. 5. Chemistry]
I. Schatell, Brian, ill. II. Title.
QD321.C695 1983 547 82-48457
ISBN 0-397-32021-3
ISBN 0-397-32022-1 (lib. bdg.)

4 5 6 7 8 9 10

1 GOOEY STUFF

Ever touch something gooey? Stick your fingers in a jar of cold cream. Make some mud pies. Feel some spit. Ick!

Gooey stuff feels soft and thick and sort of wet. You can push it around easily, but it is not very runny. It has no real shape. Gooey stuff sticks together. A blob of goo will plop off a spoon or hang in a heavy drop that slowly gets longer and longer.

Go on a goo hunt around your house. How many gooey things can you find? Here's a list to get you started: toothpaste, shampoo, syrup, ice cream, glue, chewed food.

There are many kinds of goo. Some is pleasant and creamy. Some is icky and smelly. Some goo keeps out water; other goo keeps it in. Some goo makes things slip and slide; other goo sticks things together. All living things make goo, and an important part of all living things *is* gooey stuff.

You are a goo user.
You rub it on your skin.

You spread it on bread and eat it.

You do the dishes with it.

You paste with it.

You protect your lips with it,
and lots more.

All the different kinds of gooey stuff, and all the different ways it can be used make it very interesting. Some scientists study goo. They ask questions about it. What is it made of? What does it do? How can you use it? How can you make it?

You can be a scientist and make your own study of goo. This book shows you how. Make goo you can eat. Learn how goo protects you. Discover goos that stop being gooey.

Want to play with gobs of goo? Read on.

2 GREASY GOO

Feel something greasy. Put a drop of salad oil on your fingertips and rub them together. It feels wet and slippery. Wash your hands with soap. Feel some plain water. It feels different—not so slippery.

Oil and water are two kinds of liquids that give a wet feeling to goo. But you can't always tell if a goo has either water, oil, or grease in it just by feeling it.

Here is a better test to see if a goo has grease in it. Rub a small amount of some goo into a piece of brown paper bag. Try any kind of goo—jam, peanut butter, liquid soap. Wipe off the rubbed spot with a paper towel. Now hold the brown paper up to the light. You will see the light more easily through the paper where you rubbed the goo.

Now put the paper aside and wait a half hour or so, to give any water in the goo time to dry. Grease does not dry. So if the goo has grease in it, you will still see light through the paper. If the spot is dry and you cannot see the light, then the goo has only water.

Some goos are made of both grease and water. But you can't make goo just by mixing water and grease. See for yourself.

Pour some salad oil into a jar that has a lid. Gently drop a spoonful of water into the oil. The water forms perfect balls that sink slowly to the bottom of the oil. Beautiful!

Now add three or four more spoonfuls of water to the oil. Screw the lid on the jar and shake it hard. The water and oil form a cloudy mixture as drops of water spread in the oil. Set the jar on the table and watch. Slowly the drops of oil and water separate into two layers. Which one is on top? Are you surprised?

Salad dressing is made of vinegar and oil. Vinegar contains water. You always have to shake it before you pour it on a salad. And then you have to pour quickly before it separates. Mayonnaise is salad dressing you don't have to shake. It is also made of oil and vinegar, but it contains a third substance that keeps the oil and vinegar from separating. This third substance is egg.

Make some mayonnaise and see how the egg also makes the mixture thick and gooey while it keeps the oil and vinegar mixed. You will need these things:

a one-cup measuring cup
measuring spoons
a clean jar with a screw-on lid
salad oil
vinegar
an egg

Crack the egg on the edge of the jar, and open it so the egg goes into the jar. Add three tablespoons of oil. (A tablespoon is the largest measuring spoon.) Screw on the lid and shake fifteen times. The mixture will be lemon yellow. The oil should not form a separate layer when you stop shaking.

Measure out a cup of oil in the measuring cup. Pour a little at a time into the jar. Screw on the lid and shake fifteen times each time you add a little more oil. When you have added half the oil in the measuring cup, put two tablespoons of vinegar into the jar. Again, shake hard fifteen times. Keep on adding the rest of the oil, a little at a time, shaking after each addition. The mixture will get thick and creamy. You can stop shaking it after all the oil is mixed in. To season it, add a teaspoon of salt and a half teaspoon of pepper and stir well.

Homemade mayonnaise is not as thick as mayonnaise from the store. But some people think it tastes better. Keep yours in the refrigerator.

A mixture of oil and water that is held together by a third substance is called an *emulsion.* Mayonnaise is an emulsion. Egg holds the oil and the water that is in the vinegar together. Soap will also make an emulsion of oil and water. Shake some liquid soap in a jar with water and oil, and see for yourself.

Still another emulsion is heavy cream. The fat is spread throughout the cream in very tiny drops. A substance in milk that is something like egg holds the fat and water together. You can get the fat out of heavy cream. Guess what! The fat in a cream emulsion is butter.

To make butter, pour some heavy cream in a jar so it is half full. Screw on the lid and just shake. You will have to shake so long and so hard that it's easier to make butter with some friends. That way you can take turns shaking the jar when you get tired.

At first, the cream will get thick, like whipped cream. Be patient, keep shaking. This takes about ten minutes. Then the cream will get very hard to shake. It will stick to the sides of the jar. Keep shaking. Next, you will see it begin to stick together. You can see spaces on the sides of the jar. Then it will all cling together. Suddenly it will get very easy to shake as you squeeze the buttermilk out of the butter, which sticks together in one big gob. Spread it on bread and eat it. (You can also drink the buttermilk.) It will be the best butter you ever ate. Yum!

There are other kinds of greasy goo that are not good to eat. Oil from oil wells is made into greases that make machines run more smoothly. The moving parts in cars and other machines are greased so that they slip and slide against one another. That way they don't wear out so fast.

Grease also protects a surface from air and water. Metals that are coated with grease will not rust. Some oils burn well and are used as fuel to heat homes.

These are just a few of the things greasy goo can do.

3 STICKY GOO

Some sticky goo is good to eat. Honey, syrup, jam, and jelly are all sticky and all sweet. Do you think they might be good to use as glue? Experiment and see.

Spread some sweet, sticky goo on a piece of paper. Put another piece of paper on top. Can you pull them apart easily? Leave the papers stuck together until the sweet goo dries. This may take an hour or more. Now try and pull the paper apart. How good is your glue?

Sweet, sticky goo contains sugar and water. When you stir sugar into a glass of water, the sugar disappears. But even if you can't see the sugar any longer, you know it's there because you can taste it.

The water separates the sugar into its tiniest parts. These tiny sugar parts are called *molecules*. They are so small no one has ever seen them. Different kinds of things are made of different kinds of molecules. Sugar molecules are different from water molecules. All molecules are very tiny. Scientists have ways of knowing their size without seeing them. How scientists know about all kinds of molecules, not just sugar molecules, is another story. You can read about it in other books.

Syrup and honey are sticky because they have a lot of sugar in a small amount of water. Wet sugar molecules stick together. They also stick to paper. That's why you can use sweet, sticky goo as glue.

Mix flour and water to make another kind of sticky goo. Put one tablespoon of flour in a cup. Let the hot water run until it is as hot as it will get. Stir two tablespoons of hot water into the flour. It will become a thick paste.

Spread your flour paste on some paper. Cover it with another piece. Pull them apart. Is flour paste as sticky as syrup? Put the papers together again, and let your paste dry. Now try to pull the papers apart.

Sugar molecules can join together to form long chains. Naturally, a molecule made of chains of sugar molecules is much bigger than a sugar molecule. When sugar molecules form chains they lose their sweetness and become a different substance called a *starch*. Flour is a starch. Flour molecules are too big to disappear in water. Instead, they soak up water molecules. This makes them swell. The long chains get tangled with one another. The long tangled chains will hook into a rough surface like paper. That's why this substance is sticky. When the water dries, the starch chains stay tangled and hooked into the paper's surface. Starch makes a good paste.

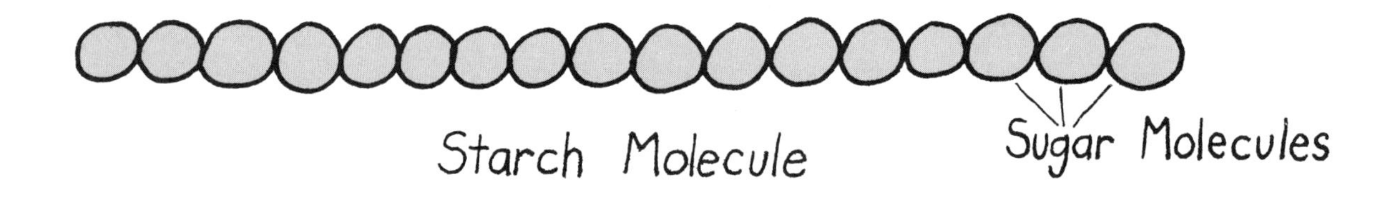

Paste and glue you buy at a store are not made of sugar or flour. These may be made of other substances that can be used to stick things together. Rubber, plastic, and animal bones are used to make glue. They all have molecules that are long chains.

When you use a paste or glue, you spread it in a thin layer between the things you want to stick together. The paste or glue dries into a thin film. You can see what these films are like.

Collect any kind of paste or glue you have around your house. Put a dab of glue on a flat plastic top, such as the one that covers a coffee can. Spread the dab into a thin circle with a toothpick. Let it dry. When it is completely dry it will not feel sticky when you touch it. Peel the dried glue off the plastic. Now feel the side that was next to the plastic.

I'll bet you never knew that sticky goo could get to be so smooth!

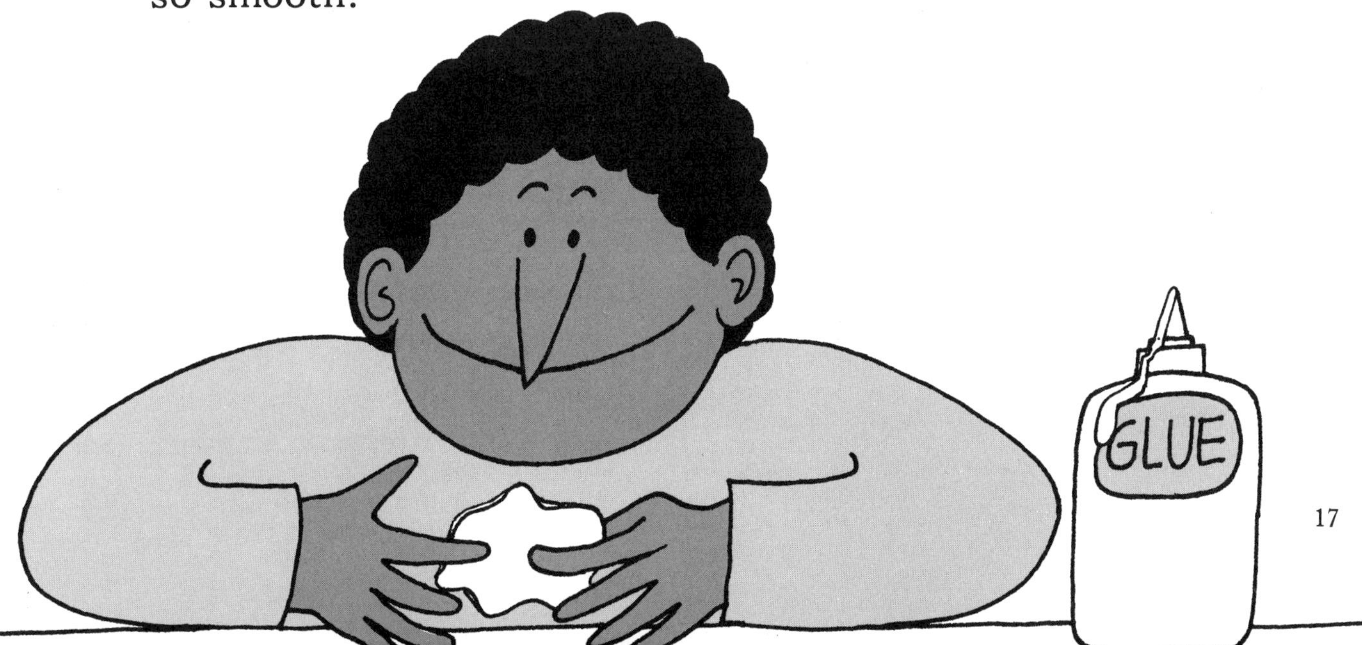

4 SLIMY GOO

Want to feel some slimy goo? Let a raw egg run through your fingers. Pick up a wet fish or hold a raw oyster. Touch a trail left on glass by a land snail.

Slimy goo feels wet and thick and a little like greasy goo and a little like sticky goo. Snails make a kind of slime

to travel on. Their slime helps them stick to a smooth surface like glass. It also helps them glide slowly along their path.

Egg white, another slimy goo, protects the yolk inside its shell. See for yourself. The next time someone is going to cook eggs, hold a raw egg and shake it hard. Then crack it open into a bowl, carefully. If you don't break the yolk when you crack the egg, the yolk will still be whole. The egg white kept the yolk in the middle of the egg and protected it from banging against the shell. The mother hen often moves her eggs. The white makes sure the yolk stays whole so that it can develop properly.

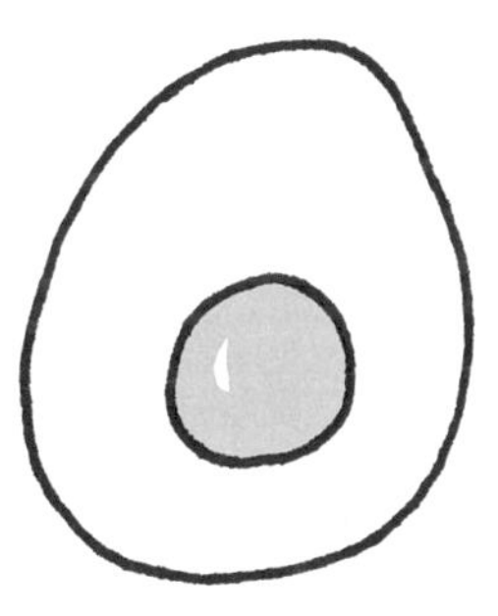

Slimy goo protects things that live in water. Many water plants in fish tanks, frogs' eggs, and fish are coated with slime. When you hold your hands in water for a long time, your fingers get to look like prunes. Water flows into your skin and makes it swell and get wrinkled. A coat of slimy goo keeps this from happening to many things that live in water. Water can't get through the slime.

Here's something you can do with slimy goo. You can blow bubbles in it. Try blowing bubbles in egg white. To do this, you have to separate the yolk from the white. This is tricky and may take some practice. The idea is to keep the yolk whole. You can scramble your mistakes. You will need these things:

two small bowls
a table knife
an egg
a jar or juice glass
a straw

Give a sharp rap with the blade of the knife to the middle of the side of the egg. It should crack cleanly. If it doesn't, give another light tap in the same place.

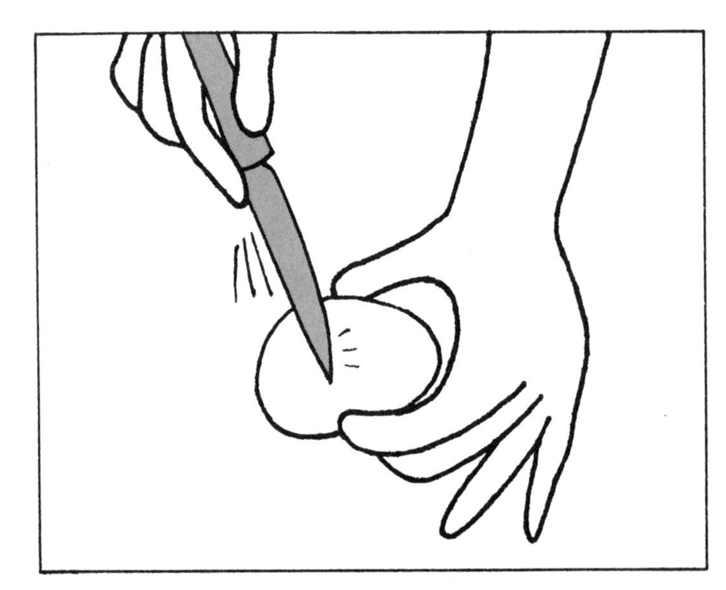

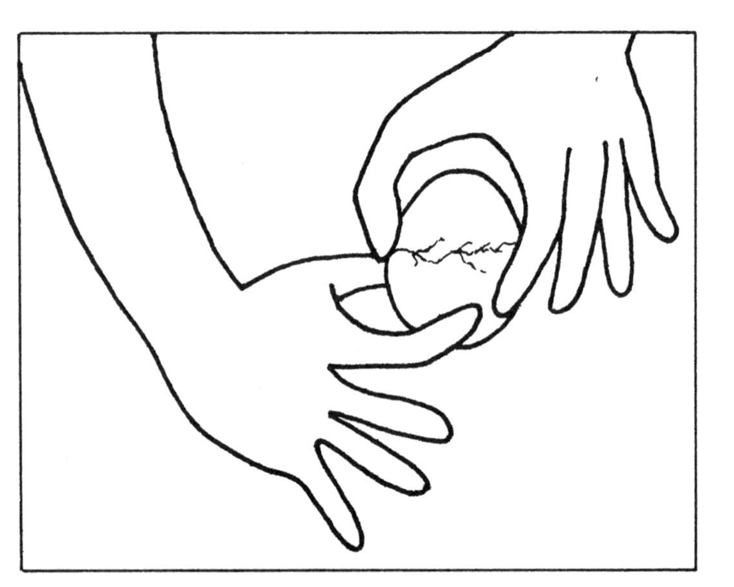

Hold the cracked egg in an upright position in your left hand over one of the bowls. Put the thumbnail of your right hand into the crack and pry off the top half of the eggshell. (Use the opposite hands if you are left-handed.) When you remove the top of the eggshell, egg white will run over your hand into the bowl. If the yolk did not break, you're halfway there. If the yolk is broken, dump the egg into the bowl and start over with a fresh egg and another bowl.

Pass the yolk from one half of the eggshell to the other over the empty bowl. Let egg white run into the bowl each time. Keep passing the yolk back and forth between eggshell halves until there is no longer any white around the yolk. Put the yolk in the other bowl.

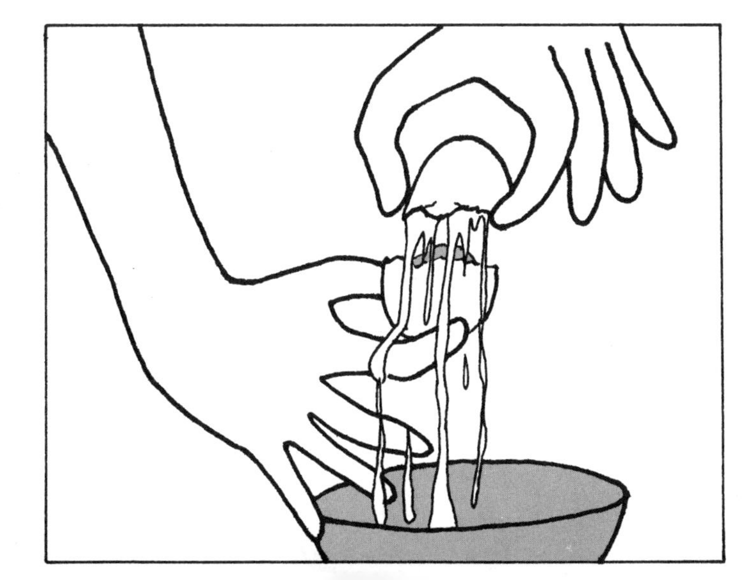

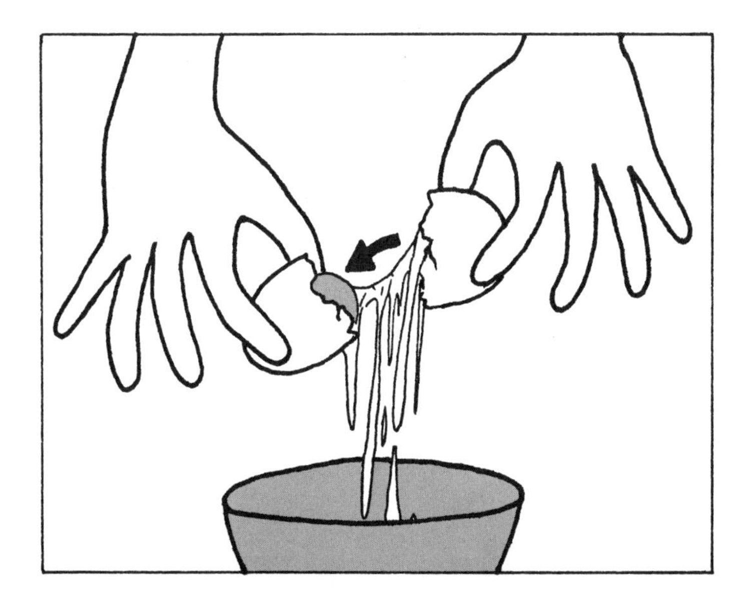

Pour your egg white into a tall, skinny olive jar or a juice glass. Stick a straw in it and blow bubbles. The egg-white foam will climb up the sides of the jar. Mix in a tablespoon of water. Blow bubbles again. When you add water, does it make better foam?

Egg white is a mixture of water and a substance called *protein.* Every living thing has protein in it. The kind of protein in egg white makes the mixture thick and slimy. The protein and water form a thin film when you blow into it that is the wall of a bubble.

Liquid soap is another slimy goo that's great for blowing bubbles. Put a teaspoon of liquid soap in another tall, skinny jar or glass. Add three tablespoons of water. Stick a straw in it and blow bubbles into the soapy water. Which foam lasts longer, egg-white or soap?

See if you can blow bubbles in the air by dipping a straw into soapy water and then blowing gently. Do the same with egg white. Can you get an egg-white bubble to float? It's harder than floating soap bubbles.

See if other slimy goo can be made into bubbles. Try gelatin dessert or cold, jellied chicken soup.

5 WARM GOO

Some goo isn't gooey until you warm it up. Before you heat it, it is a solid. But when you add heat, it can get gooey and even runny. When it cools, the runny fluid becomes a solid again.

Chocolate candy, wax, and cheese all get gooey when you heat them. Heat makes them *melt*. Here's the idea behind melting.

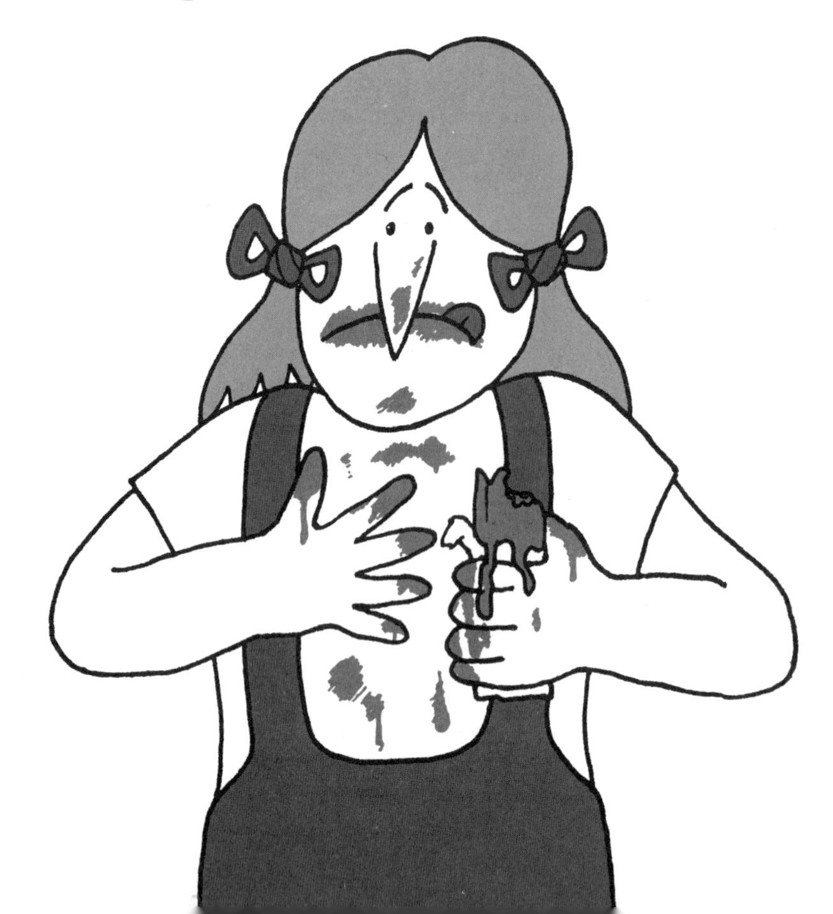

Before something melts, it is a solid. All the molecules hold each other in place. When you add heat, the molecules start shaking. The hotter it gets, the harder they shake. Pretty soon they shake free of one another. They start tumbling over one another. The stuff is no longer a solid. It has melted and become a fluid that can flow from one place to another.

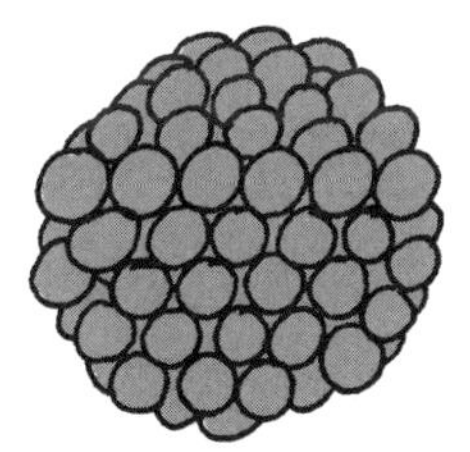

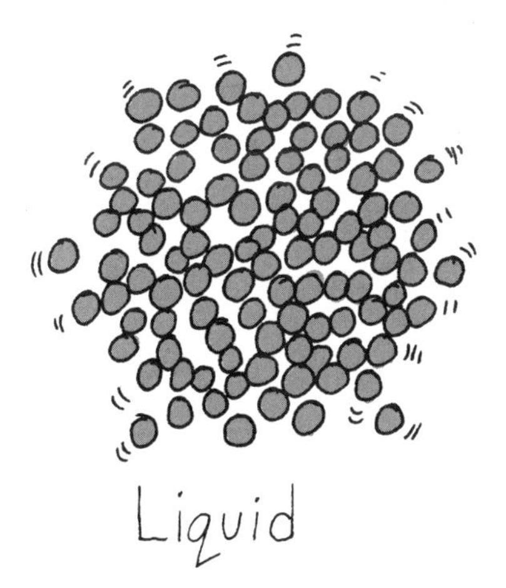

When the melted fluid cools, the molecules move more and more slowly. They stop tumbling past one another and take new fixed places. They become a solid, holding one another together again.

See it all happen for yourself. Warm up some goo and then watch it cool. You will need these things:

- 3 aluminum muffin cups
- an old candle
- a piece of paper
- a hammer
- some chocolate candy
- some cheese (pizza cheese gets runny and stringy)
- a lamp
- a pot holder

Put some chocolate candy in one muffin cup. Put some cheese in another. Put the candle on a piece of paper. Hammer it to break it up into pieces. Put the smallest pieces in the third muffin cup.

Check with a grown-up before you do the next part. Things can get hot. Put each muffin cup on top of an unlit light bulb in a lamp. Make sure each cup is not going to fall. Wedge them between the wires holding the lampshade to keep them in place. Turn on the lamp. Watch as your stuff starts melting.

When your goo has completely melted, take it off the light bulb with the pot holder. The muffin cups will not be hot enough to burn you, but they will be quite warm. Be careful not to spill your melted goo! It's very hard to clean up. Set the cups down on a table and watch as they cool.

Did the shape of your goo change after it melted? Which part cools first, the outside or the center? When your goo is cool and hard, try to remove it from the muffin cups. Turn each cup upside down and tap the bottom. Does it fall out? What shape does it have?

Plastic wrap and plastic bags are made of stuff that is like wax. It melts when you heat it. It is called *polyethylene*. Polyethylene can be made into many different shapes when it is warm. A thin layer becomes sheets used for plastic wrap and bags. Toys are made from polyethylene that has been poured into molds. The polyethylene takes the shape of the molds much as the wax took the shape of the muffin cup.

Put a plastic bag in an aluminum muffin cup. (You may have to cut it up into small pieces.) Put the cup on a light bulb. Watch what happens.

If you put plastic pieces in a muffin cup and they don't melt, you do not have polyethylene. You have a different kind of plastic.

Now you know why you don't want to rest a plastic bag on the toaster.

6 GOO FROM YOU

The outside of your body, your skin, is the driest part of you. Inside your body it is wet and gooey. The wetness comes from water.

Water is in every part of you. It is in liquids like tears, sweat, blood, saliva, and urine. It is also in your more solid parts like muscles, and bones, and your brain. In fact, you have more water in you than any single other substance. If you weigh sixty pounds, thirty-six pounds of you is the water in your body. That's more than half of you! No wonder your insides are gooey!

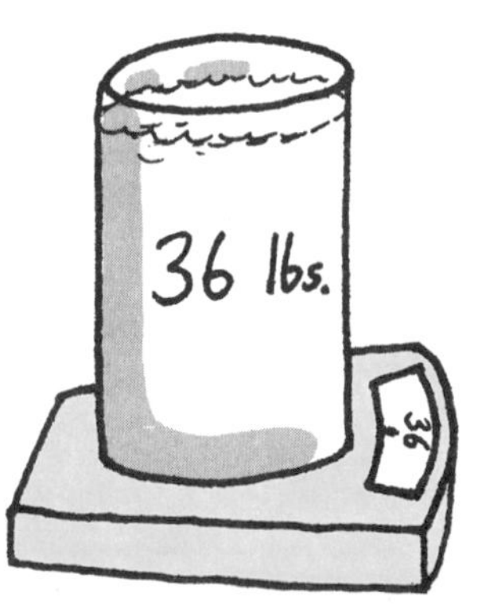

You need to drink water to keep you alive. Without water, you could not digest your food. The food molecules could not reach the different parts of your body for energy and growth. Oxygen molecules you breathe could not reach all the places they are needed. You could not get rid of your wastes. Without water you could not feel or see or move. You would not grow. You could not live.

Most of the water you take into your body has become a part of you. It is inside the smallest living parts of you, called *cells*. You can see cells under a microscope. Here's a picture of some of your different kinds of cells. Water found in cells is not runny or watery. It is more like jelly.

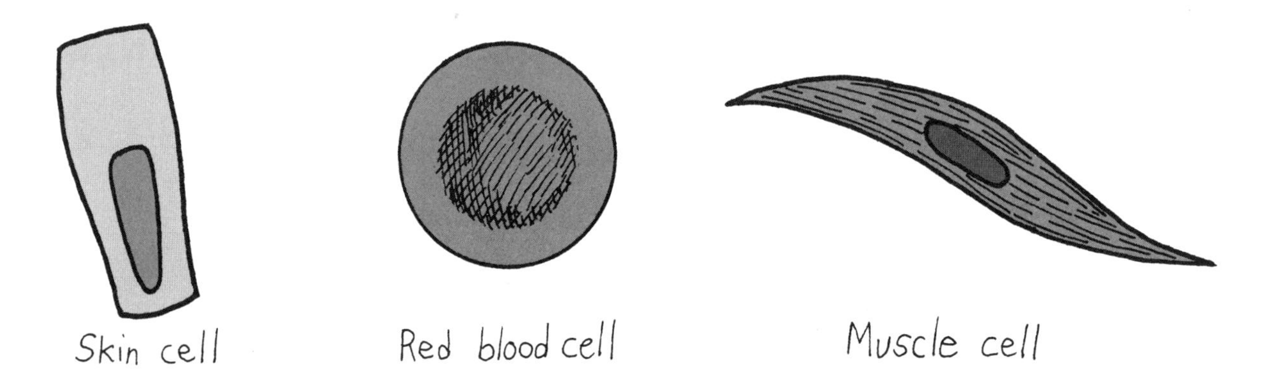

Your body needs to take in water because your body also makes fluids. The liquids your body makes that seem most like water all pass out of cells. Tears, sweat, and urine are not pure water. They all contain salt. Urine contains other wastes that would poison you if they were not carried out of your body.

Some substances from your body contain more than salt. Molecules in these liquids make them thicker and less runny than tears. These are the goos I want to tell you about in this chapter.

Saliva is stickier and slimier than plain water. It needs to be because of the job it does. When you chew your food, you mix saliva with it. Saliva helps the food move more smoothly down the long tube leading to your stomach. It also holds the chewed food together, making sure it all reaches your stomach and doesn't collect in your mouth or in the tube.

Saliva also gets food ready for your stomach. It starts breaking down some kinds of food so that it can be taken into cells during digestion. See for yourself. Chew a plain soda cracker, one without salt. Hold it in your mouth without swallowing for five minutes. Now taste the chewed cracker with the tip of your tongue. It will taste sweet. Saliva changes starch in crackers and potatoes and other foods into sugar. Sugar molecules can be taken into cells. Starch molecules are too big to get in.

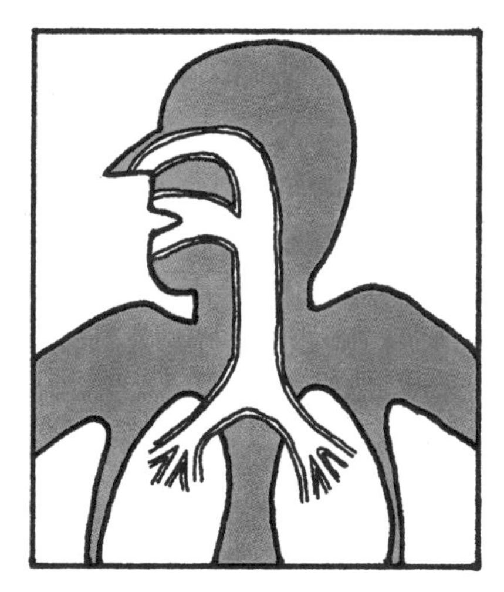

The inside of your nose and the tubes that lead to your lungs are lined with another kind of goo called *mucus.* Thick, slimy mucus protects you from dust and germs in the air. It acts like flypaper. Dust and germs get stuck in it. Sometimes germs get past the mucus into your cells. Then you catch a cold. Your cells react by making lots of mucus and giving you a runny nose. The extra mucus is trying to flush the germs out of your head.

If you have ever had a cut that became infected, you've seen another kind of goo, namely *pus.* Pus is a yellowish white, thick, creamy material. It shows that your body has been fighting a war with the germs that infect you. Here's how it forms.

When you cut yourself, blood comes out of the cut. If you do not keep the cut clean, germs can get into the cut. When this happens blood rushes to the wound to fight the germs. The area around the cut becomes swollen and red and sore. The blood contains special cells, called *white cells*, that surround the germs and take them inside.

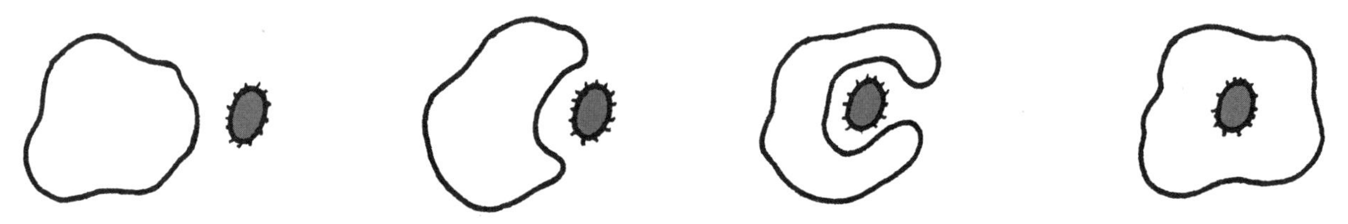

Many white cells die from eating germs. Dead white cells mixed with some of the liquid part of the blood become pus. If you clean a cut right away, your white cells won't have to fight germs and there will be no infection. White cells are one way your body protects you against germs. It makes extra white cells when there is a danger of infection.

Some people may think that goo from you is disgusting. But science shows how gooey stuff in your body does important jobs for you. If scientists thought goo was disgusting, they would not study it. We would not know what goo can do. Now you know some of those things. Yea!

GOOS THAT DO FOR YOU

FATS AND OILS

Scientific name: Lipids

What They Do and Don't Do

Do float on water
Do feel greasy and slippery
Don't mix with water

Some Things You Do with Them

Spread on bread (butter and oleo)
Make salad dressing (salad oil and mayonnaise)
Burn for heat and light (fuel oils)
Grease your bicycle and other machines

SUGARS

Scientific name: monosaccharides, disaccharides

What They Do

Do taste sweet
Do dissolve in water
Do give quick energy as food

Some Things You Do with Them

Make syrup, jam, jelly, candy, molasses

STARCHES

Scientific name: polysaccharides

What They Do and Don't Do

Don't dissolve in water
Don't taste sweet
Do absorb water and swell

Some Things You Do with Them

Thicken gravy
Make paste
Make bread and cakes
Starch clothes to make them stiff

SLIMY GOOS

Scientific names: egg albumin, mucopolysaccharides

What They Do

Do mix when stirred gently with water
Do foam when beaten
Albumin sets and gets white when heated

Some Things You Do with Them

Chew your food (saliva is a mucopolysaccharide)

SLIMY GOOS

Some Things You Do with Them (cont.)

- Protect the inside of your nose (mucus is a mucopolysaccharide)
- Scramble eggs (egg albumin gets solid when heated)

WAXES

Scientific name: Esters of higher fatty acids

What They Do and Don't Do

- Don't mix with water
- Do melt at temperatures warmer than your body but cooler than boiling water
- Do become shiny when rubbed

Some Things You Do with Them

- Polish shoes, floors, and cars
- Use as lipstick
- Protect as lipstick
- Burn as candles

THERMOPLASTICS

Scientific name: There are many different kinds, polyethylene is a good example

What They Do and Don't Do

- Do feel waxy
- Don't break easily
- Do come back to shape after being squeezed
- Do melt like wax

Some Things You Do with Them

- Wrap food
- Use toys made from them
- Use bowls, squeeze bottles, and many other products made from thermoplastics